TECH KNOW SOLAR PV SYSTEM

AN OVERVIEW OF SOLAR PV TECHNOLOGY

FIRST EDITION

BY PRASUN BARUA

TABLE OF CONTENTS

INTRODUCTION

Welcome to the TECH KNOW SOLAR PV SYSTEM! This book contains various types of topics on solar PV technology. This is an overview of the Solar PV technology. Solar PV (Photovoltaic) technology is one of the significant technologies contributing in solar PV industry. This environmentally friendly technology helps to reduce carbon emission and keep our environment clean and healthy. Its operation and maintenance cost are comparatively less. By using this technology, people also get benefited both economically and socially. After reading this book, you will know about the solar PV technology.

Solar PV system components, how solar panels are made, how to choose a solar panel, how to size components for the system, how to design the solar PV system, how to assess and select a site for solar PV project, how to install solar panels, why diodes are used in solar panels, grounding solar PV system, how to build a solar power plant, how to maintain & troubleshoot the system and the challenges is solar power generation are described in this book. This the first edition of the book. It will be great pleasure if this book helps you to know about solar PV technology. Thanks for reading the book.

CHAPTER-1: What is a solar cell and how does it work?

Solar cell is an electrical device which converts the light energy directly into electricity utilizing photovoltaic effect. It is also defined as the form of photoelectric cell having electrical characteristics like current, voltage and resistance.

There are at least two semiconductor layers in solar cells. One layer contains a positive charge and the other layer contains a negative charge. A typical silicon solar cell consists of a thin wafer having phosphorus doped (N-type) ultra-thin silicon layer on the top of boron-doped (P-type) thicker silicon layer. When these two materials are connected with each other, a junction is formed which is called P-N junction. As a result, an electrical field is created near the top surface of the cell. When these two layers are connected to an external load, the electrons flow through the circuit generates electricity.

Sunlight consists of small particles of solar energy called photons. When sunlight strikes on the surface of solar cell, many of the photons are reflected and absorbed by the solar cell. Electrons are released from the negative layer of semiconductor material, when enough photons are absorbed by this layer of the solar cell. These electrons naturally move into the positive layer and create a voltage differential.

Under open circuit, no load conditions, a typical silicon solar cell generates about 0.5 – 0.6 volt DC (Direct Current). The output of a solar cell depends on its surface area (size) and efficiency. It is also proportional to the light intensity of the sun striking the surface of the cell. For example, under peak sunlight conditions, a typical commercial solar cell with a surface area of 160 square centimeters will approximately generate peak power of 2 watts. If the intensity of the sunlight is 40 percent of peak, then the cell would generate approximately 0.8 watts. In order to increase the output power, cells are combined in a weather tight package which is called a solar module. In order to create the desired voltage and amperage, these modules (from one to several thousand) are then wired up in series and parallel with each other. It is called a solar array.

The semiconductor material silicon which is primarily used for the manufacturing process of solar cells is naturally available. Due to the natural availability of silicon and the practically unlimited resource in the sun, solar cells are very environmentally friendly. Solar cells burn no fuel and have absolutely no moving parts which makes them virtually maintenance free, silent and clean.

CHAPTER-2: How solar panels are made?

A solar PV (photovoltaic) module contains solar cells, glass, EVA, back sheet and frame. Three types of solar panels are available in the market. They are:

➢ Mono-crystalline solar panels
➢ Poly-crystalline solar panels
➢ Thin film solar panels

Therefore, various types of materials are used for manufacturing at cell structure level. They are - mono silicon, poly silicon or amorphous silicon. Mono and Poly crystalline cells have almost similar manufacturing process. For producing a crystalline solar panel following steps are followed:

First Step: Sand

Here sands are used as a raw material. Most solar panels are made of silicon, which is the main component in natural beach sand. Silicon is plentifully available which is the second most available element on Earth. However, converting sand into high grade silicon is a high cost energy intensive process. Pure silicon is produced from quartz sand in an arc furnace at very high temperatures.

Second Step: Ingots

Basically, the silicon is collected in the form of solid rocks. Hundreds of these rocks are being melted together at very high temperatures in order to form ingots in the shape of a cylinder. A steel, cylindrical furnace is used for forming desired shape. All atoms need to be perfectly aligned in the desired structure and orientation during melting process. For providing the silicone positive electrical polarity, Boron is added to the process

Mono crystalline cells are manufactured from a single crystal of silicon. Mono Silicon has higher efficiency in converting solar energy into electricity, therefore the price of mono crystalline panels is comparatively higher.

Poly silicon cells are made from melting several silicon crystals together. After the ingot has cooled down, grinding and polishing are being performed, leaving the ingot with flat sides.

Third Step: Wafers

In this step, wafers are used during manufacturing process. The silicon ingot is sliced into thin disks, also called wafers. A wire saw is used for precision cutting. The thinness of the wafer is similar to that of a piece of paper. As pure silicon is shiny, it can reflect the sunlight. An anti-reflective coating is put on the silicon wafer for reducing the amount of sunlight lost.

Fourth Step: Solar cells

Treating each wafer, metal conductors are added on each surface. The conductors provide the wafer a grid-like matrix on the surface. This confirms the conversion of solar energy into electricity. The coating will ease the absorption of sunlight, rather than reflecting it. In an oven-like chamber, phosphorous is being diffused in a thin layer over the surface of the wafers. This will charge the surface with a negative electrical orientation. Boron and phosphorous combination will provide the positive - negative junction, which is very important for the proper function of the PV cell.

Fifth Step: From Solar Cell to Solar Panel

In this step, using metal connectors, the solar cells are soldered together to connect the cells. Solar panels are made of solar cells integrated together in a matrix-like structure. The current standard offering in the market are:

- 48 cell panels - For small residential roofs.
- 60-cell panels - The standard size.
- 72-cell panels - For large-scale solar power plant.

After putting the cells together, a thin layer (about 6-7 mm) of glass is added on the front side, facing the sun. Highly durable, polymer-based material is used to make the back sheet. This will protect solar panel entering water, soil and other materials from the back. For enabling connections inside the module, the junction box is added.

After assembling the frame, it all comes together. The frame protects the panel from impact and weather. The use of a frame will also allow the mounting of the panel in a variety of ways, for example with mounting clamps. EVA (ethylene vinyl acetate) is the glue which binds everything together. It is crucial that the quality of the encapsulation is high so it doesn't damage the cells under harsh weather conditions.

Sixth Step: Testing the Modules

For ensuring expected performance of the cells, testing is carried in this step. perform as expected. STC (Standard Test Conditions) are used as a reference point. The panel is put in a flash tester at the manufacturing facility. The tester will deliver the equivalent of 1000W/m2 irradiance, 25°C cell temperature and an air mass of 1.5g. Electrical parameters are written down and these results can be found on the technical specification sheet of every panel. The ratings will reveal the power output, efficiency, voltage, current, impact and temperature tolerance.

Besides STC, every manufacturer uses NOCT (nominal operating cell temperature). The parameters used are more close to 'real life' scenario: open-circuit module operation temperature at 800W/m2 irradiance, 20°C ambient temperature, 1m/s wind speed. The ratings at NOCT can be found on the technical specification sheet.

Before shipping the module to homes or businesses, cleaning and inspection are done which is the final step of the production.

The aim of the research and development in the solar energy industry is to reduce the cost of solar panels and increase the efficiency. The solar panel manufacturing industry is becoming more viable and is predicted to become more popular than conventional sources of energy like fossil fuels.

CHAPTER-3: Why are diodes used in solar panels?

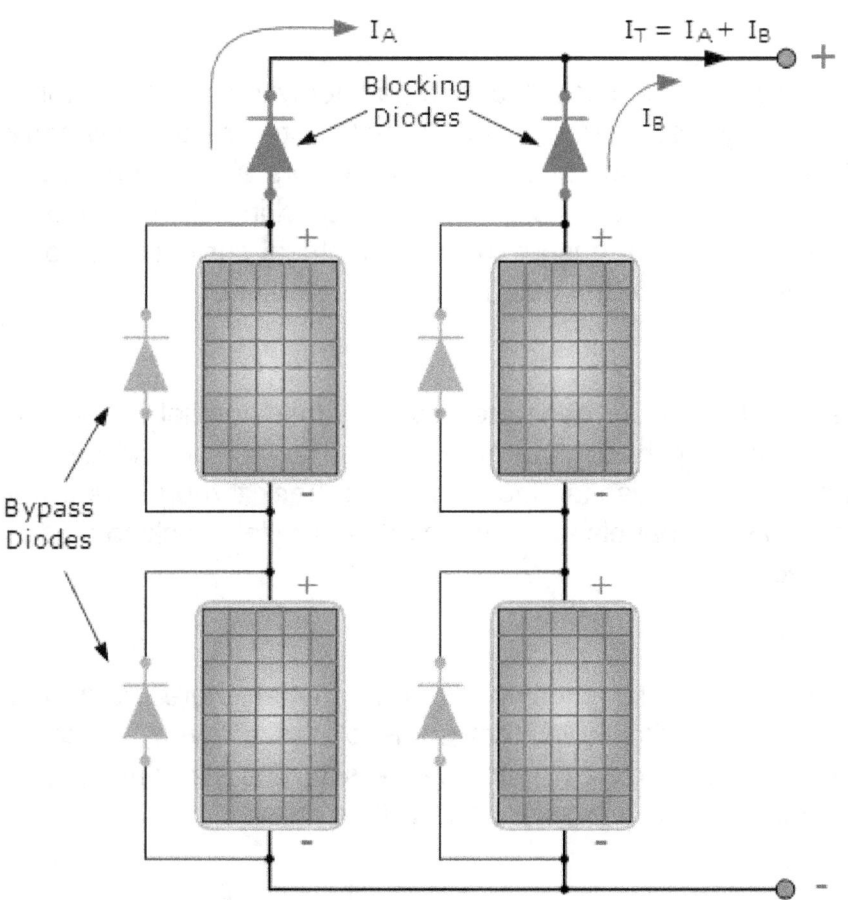

Diodes are two terminal electronic component which allow current to flow in one direction. This diode can be used to block the flow of electric current from other parts of an electrical circuit. These types of silicon diodes are basically known as Blocking Diodes during the usage with a solar panel. Bypass Diodes are connected in parallel with either a single or a number of solar cells to prevent the solar cells overheating by sunlight. By providing a current path around the bad cell, these types of diodes also protect the partially shaded solar cells from burning out. Blocking diodes are used differently than bypass diodes.

Bypass diodes in solar panels are connected in "parallel" with a solar cell or panel to shunt the current around it, whereas blocking diodes are connected in "series" with the PV panels to prevent current flowing back into them. Therefore, blocking diodes are different than bypass diodes. These types of diodes are physically the same in most cases. In order to serve various purposes, they are used in different way.

Usage of Bypass Diodes in Solar Photovoltaic Arrays

Here, diodes with green color are "bypass diodes", one in parallel with each solar panel to provide a low resistance path. Bypass diodes in solar panels and arrays safely carry this short circuit current. On the other hand, diodes with red colors are known as the "blocking diodes", one in series with each series branch. These blocking diodes are also known as a series diode or isolation diode, ensure that the electrical current only flows in one direction "OUT" of the series array to the external load, controller or batteries.

The reason for this is to prevent the current generated by the other parallel connected solar panels in the same array flowing back through a shaded solar cell and also to prevent the fully charged batteries from discharging or draining back through the array at night. Therefore, when several solar panels are connected in parallel, blocking diodes are used in each parallel connected branch.

Blocking diodes are basically used in solar photovoltaic arrays when there are two or more parallel branches or there is a possibility that some of the array will become partially shaded during the day as the sun moves across the sky. The size and type of blocking diode used depends upon the type of solar photovoltaic array.

Two types of diodes are available as bypass diodes in solar panels and arrays. One is the PN-junction silicon diode and another is the Schottky barrier diode. Both diodes are available with a wide range of current ratings. Forward voltage drop of the Schottky barrier diode is about 0.4 volts/ On the other hand, the PN junction silicon diodes have the voltage drop of 0.7 volt for a silicon device.

This lower voltage drop helps to save one full solar cell in each series branch of the solar photovoltaic array. Because of dissipating less power in the blocking diode, the array becomes more efficient. During manufacturing solar panel, most manufacturers add both bypass and blocking diodes in their solar panels.

CHAPTER-4: How to assess and select a site for solar PV project?

In the process of selection, design and installation of an appropriate solar PV system for your home or business, a professional site assessment is a crucial part. Cost and power output of your potential PV System significantly depend on it. The output of a PV module is directly proportional to the amount of sunlight strikes on it.

A proper site assessment significantly contributes to determine and identify a number of factors: the site capacity of generating energy, shading issues, how much revenue will be generated from the power plant, return on investment (ROI), rebates and incentives, structural and electrical concerns. The site assessment also significantly contributes to save money in basic energy efficiency improvements and more lucrative methods to organize a solar project on-site. After reviewing the report, you will be able to make well-versed decisions about your project. The site assessment report will help you to bring your goals, budget and energy needs together with the unique solar opportunities at your location. Every site is different and needs evaluation specific to the site.

Electricity is produced from PV Modules produce when photons on solar cells and knock available electrons loose and into motion. Photons are small packets of energy contained in sunlight. When fewer photons strike on the solar cell, for example due to poor orientation or haze, fewer electrons are put into motion. As a result, little amount of electricity is produced. But if there is shade in the site even with little amount, it can cause shutting down the production completely in some cases.

Modules with built-in bypass diodes contribute to minimize the effects of partial shading. But, even a row of cells with shade can disable the module. Impact of shading requires careful site planning and design considerations for solar PV arrays. Whether it is a neighbor's multistory home or trees on your property, most sites should be considered at least some shade. While wide-open, dawn-to-dusk exposure is ideal, PV system designers generally shoot for a shade-free solar window from 9 a.m. to 3 p.m. ("solar time" for all days/months of the year). Majority of solar radiation is available during these hours. However, it may be affected by local climate variations. For example, in some locations, early morning fog can shift the "prime" solar window toward sunnier afternoon hours.

If forecasting shading throughout the year has been done by sight alone from various barriers like tall trees, nearby buildings, roof dormers and even chimney, then it can be challenging which requires many observations over the course of the year. But some tools like the Solar Pathfinder, the Acme Solar Site Evaluation Tool and the SunEye, can assist you to assess shading on your site throughout the year quickly with one site visit. Each tool has different technique and price. But, the job can be done by all these tools. They can be used at a proposed array location for the evaluation.

Historical solar radiation and weather data for your latitude and longitude and the constantly changing sun elevation angle are considered by these tools. To provide additional data for accurate shade compensation calculations due to tall trees, nearby buildings etc. digital photos are taken at the site. Depending on the solar panel tilt (up and down angle), azimuth (right and left orientation) and whether a tracking system is to be employed, some modifications are done. Wire runs, connections, fuses and breakers, inefficiencies of the inverter and snow shading etc. are also needed to be considered for the final output of power production of potential solar PV system.

CHAPTER-5: How to design a solar PV system?

In order to design a solar PV system, you need to follow following steps:

1. Determine the amount of electricity (kWh) consumption and the electricity rate

It is important to determine the amount of electricity consumption per year for better understanding the economics of installing a solar system. On your utility bill, your electricity usage is summarized wherein natural gas usage should be excluded. As the path of the sun varies with the seasons, a solar PV system would offset your electricity demands mostly in the summer months. It should be also considered that most utility companies have a tier rate structure based on consumption patterns and state mandates. A small PV system significantly helps to reduce your electricity demands for moving you to a different tier, minimize your rates and saves your money.

2. Determine your available roof space (sq-ft)

Knowing your available roof space is crucial for determining the quantity of solar panels you can actually use. Based on the size of your house, you should draw a diagram of your roof on a piece of paper. You can also use the free tool Google SketchUp or a free CAD program DraftSight. After determining the size of your roof, it is important to know how much of your roof faces towards the equator (South in the northern hemisphere [USA, Canada, etc.], north in the southern hemisphere [Australia, South America etc.]). Solar irradiance is strongest coming from the equator and you will want to maximize your system efficiency.

3. Determine your solar radiation data and calculate the amount of energy you could produce

In this case, you should use the official government statistics available on your area. PV Watts' or Solmetric's solar calculators also can be used.

4. Determine shading issues

You should be careful if there are obstacles like trees or buildings in the vicinity of your roof which may cause shade. Shading decreases the output of your system and may also damage the solar cells. Solar Pathfinder or Solmetric Sun Eye should be used for detailed shading analysis.

5. Choose an appropriate solar panel

Various types of solar panels are available in the market. Therefore, it is very important to choose an appropriate solar panel carefully. The solar rating and the size (surface area) of the module are the most important factors. Although the solar ratings might be the same for some modules, the voltage and current output might vary. Therefore, it is advised to choose a lower voltage module for smaller projects having capacity less than 4 kW. Factors like financial limitations and color are also should be considered.

6. Choose an inverter

It is very crucial to choose an inverter out of the many available in the market. You can consider to use micro inverters which are easy to install and can generate AC power directly from the solar panel. If you want to use a central inverter, you should use the string sizing tools. The most important factors for choosing appropriate size inverter are:

> ➤ The number of strings in the system.
> ➤ The maximum input current.
> ➤ The voltage on a string.
> ➤ The minimum ambient temperature during day light time when the system is supposed to run.

➤ The maximum ambient temperature of the location.

7. Select a racking system to mount your solar PV system on your roof

It should be remembered that the racking system will consist 10% to 25% of the total costs. You should decide carefully which racking system is most suitable for your budget and for your roof. The racking system is one of the key components which protects both the roof and the modules and may last for at least 20 years. A minor mistake with the racking can permanently damage the roof or injure people or property.

Different types of racking and mounting systems are available in the market for roof mounted solar systems. Check the manufacturer's websites, read the installation manual. Be aware of the plan and codes how you are going to mount the racking system on the roof. Know about flashing solutions.

Using an inexpensive racking solution does not mean lower quality, it might require more effort to install it or the materials may be different. Additionally, as the solar PV field matures unique racking solutions, such as hanging panels vertically along an exterior wall, become more applicable.

8. Know your local and state incentives; calculate how much money you could save

You should check the websites of your local and state incentives provider. There are also a number of other financial calculators available online. Most importantly, you should estimate your electricity generation and compare it to your utility rate. It should be remembered that different tier rates might apply based on your energy consumption. Know about solar programs of your local utility company and their applications process. Implement the benefit of tax cuts and incentives.

9. Prepare your documents

There are different permitting processes in different location. In order to receive a construction permit or to apply for an incentive, you must obtain the equipments' (solar panel, inverter, racking) data sheets, submit a roof drawing and electrical drawing. You can draw those on your own using the programs mentioned above, or ask a professional to draw them for you. In order to determine the wire sizes, voltage drops, AC and DC disconnect, electrical background and knowledge is required. If you are not confident enough to do so, you should contact an electrical engineer.

CHAPTER-6: How to choose a solar panel?

 Solar panel is one of the important elements in solar photovoltaic system. Choosing the correct solar panel is crucial factor. You need to determine the right wattage of a solar panel. The size of the solar panel in watts directly affect the cost which is basically priced in dollars per watt. Watts are related to the output of each module. For example, a 50 watt panel installed under ideal condition will generate 50 watt-hours of electricity each hour and a 100 watt panel will generate 100 watt-hours each hour in each day. So, you can expect to pay double the price for 100 watt panel compare to 50 watt panel.

There are three main stages in solar photovoltaic system. They are:

➢ Generating power by the solar panel.
➢ Storing power by the battery.
➢ Using the power.

In order to determine the wattage correctly, you need to size the panel according to your required power usage. Battery stores power to be used later on. To maintain the constant level power storage in the battery, the solar panel needs to supply the battery same amount of outgoing power.

The cost of solar panel depends on the size (in watts), the physical size, quality of materials, warranty period, brand and certification of the solar panel. The price also depends on how many solar panels you are purchasing as part of full system package.

In general, purchasing large amount of solar panels will cost less per unit. But, choosing solar panels based on only the price is not a wise decision. Because, it may not fit the area where you want to install, not having required certifications to qualify for government subsidies and lack of solid warranty period.

Besides cost, it is important to consider the manufacturing process and materials used in the solar panel. Since all panel manufactures are not the same, therefore you should consider other factors before the decision of purchasing panels. They are:

> You need to consider the tolerance rate of solar panel. For example, you have purchased a panel mentioned 100 watts in the "nameplate". But, in reality it will be 95 watts only because of quality control issues. Therefore, a positive tolerance rate is crucial. That is, under standard condition, a panel of 100 watts will not only generates 100 watts, but also performs more effectively.
> There is a significant impact on temperature co efficient of solar panel. It is better having less percentage per degree Celsius. The price of solar panel with less percentage of temperature co efficiency is comparatively higher.
> Conversion efficiency of solar panel is also vital. It determines how much power is generated by solar panel during the conversion of light into electrical energy. For example, you have purchased two panels with same price. But, one has higher conversion efficiency than other, then it provides better value for money with correct efficiency.
> In certain climate condition, Potential Induced Degradation (PID) can be caused by stray currents triggered which is the reason for substantial power loss in the panel. Solar panel with little or no PID is considered as good.
> Embodied energy of the solar panel is another crucial factor. Here, panel's initial energy intensive production is compared with the time of pay back producing more energy.
> After installing solar panels, Light Induced Degradation (LID) may occur within few months which decreases the amount of power produced in the module. If there is little or no LID in the panel, then the panel is considered as good.
> Based on your installation application, the best type of solar cell will vary for you. There are three major types of solar cell remain. They are mono-crystalline, poly-crystalline (or multi-crystalline) and amorphous (or thin-film) silicon. High efficiency and good heat tolerance characteristics remain in mono-crystalline silicon. Due to recent development in poly-crystalline panel technology, it has been observed that panels with poly-crystalline silicon are equal to or better than many mono-crystalline in terms of heat tolerance, size and efficiency. Least amount of silicons are used in amorphous (or thin-film) silicon. Typically, thin-film contains least efficient solar cells. At present, some panel manufacturers are producing thin-film module with highest conversion efficiency.

CHAPTER-7: How to size battery for solar panel?

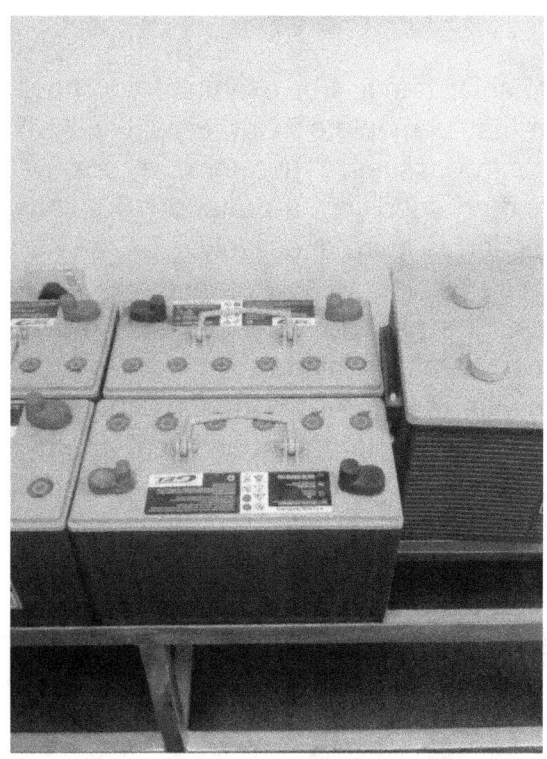

While designing a solar photovoltaic off-grid system, you need to determine required batteries. A little math is necessary to size batteries. An off-grid battery bank should be designed in such a way that the system is not only sufficient enough to supply necessary power during cloudy days but also small enough to be charged by your solar panels. First of all, you need to decide how much storage you want to provide by your battery bank. It is familiar as "days of autonomy" which is based on your expected number of days power provided by the system without receiving an input charge from the solar array. Beside determining days of autonomy, you also need to consider the critical situation and usage pattern of your application. If you decide to install a system for a weekend home, you need to consider a larger battery bank for charging and storing energy all week. But, if you want to add a photovoltaic array to a generator-based system as supplement, then you need to undersized your battery bank slightly as the generator can be operated based on recharging needed.

Rated Capacity of Battery

Basically, the capacity of a battery is specified as Ampere-Hour (AH) along with some specific standard hour reference like twenty or ten hours. For example, you have a battery which is rated at 200 Ampere-Hours and specified as a 20 hour reference. This means the battery is fully charged and will deliver a current of 10 amperes for 20 hours. If the discharge current decreases then the capacity will increase.

Depth of Discharge

The percentage of the rated battery capacity that is withdrawn from the battery is called depth of discharge. The withstand discharge capability of a battery depends on its construction. Batteries are specified by two commonly use terms. They are deep cycle and shallow cycle. Shallow-cycle batteries are comparatively cheap, lighter and have a short lifetime. Stand-alone photovoltaic systems should use deep cycle battery. Deep cycle batteries contain thicker plates and withstand up to 80% daily discharges of their rated capacity. These type of batteries are flooded electrolyte. That is the plates of batteries are covered with the electrolyte. In order to keep the plates fully covered, it is required to monitor the level of fluid and add distilled water added periodically.

Effect of Temperature

Batteries are very temperature sensitive. You cannot expect as much energy out of a cold battery as a warm one. Though, you can get more than rated capacity from a hot battery, operation at hot temperatures will decrease the lifetime of battery. It is recommended to keep your batteries near room temperature. In order to optimize the charging cycle at various temperatures and increase the lifetime of your battery, charge controllers can be purchased with a temperature compensation option.

Lifetime of Battery

It is quite difficult to predict the lifetime of any battery absolutely. Because it depends on various factors like depth of discharge, charging and discharging rate, number of cycles and operating at extreme temperature. It is quite exceptional for a lead-acid battery to last longer than fifteen years in a photovoltaic system. But, usually it can last up to eight years.

Maintenance of Battery

Periodic maintenance is required for batteries. In order to ensure that connections are tight and there is no indication of overcharging, it is necessary to check sealed battery periodically. For flooded batteries, it is required to maintain the plates well above the electrolyte level, check the voltage and specific gravity of the cells for consistent values. If the reading shows a large variation, then it may indicate cell problems. It is necessary to check the specific gravity of the cells by a hydrometer before the onset of winter particularly. The electrolyte in lead-acid batteries may freeze in cold environment. The freezing temperature is a function of a battery state of charge. The electrolyte becomes water and battery may freeze if it is completely discharged.

Sizing the Battery

The recommended battery type for using in solar photovoltaic system is deep cycle battery. Deep cycle battery can discharge to low energy level. It also can be recharged rapidly. In order to operate at night and cloudy days, the battery should be large enough to store sufficient energy. The size of battery is determined as follows:

> Determine the total Watt-Hours/day consumed by appliances.
> Divide the total Watt-Hours/day consumed by the battery loss factor (typically, it is 0.85).
> Divide the answer you got in item 2 by the depth of discharge of battery (typically it is 0.6).
> Divide the answer you got in item 3 by the Nominal Battery Voltage.
> Multiply the answer you got in item 4 with days of autonomy (the number of days required for the system to operate by solar panels when there is no power generated) to get the required capacity of the deep-cycle battery in Ampere-Hour (AH).

Battery Capacity (AH) = (Total Watt-Hours/day consumed by appliances x days of autonomy) / (battery loss factor x depth of discharge x nominal battery voltage)

CHAPTER-8: How to size wire for solar panels?

It is very important to size wires correctly for reaching energy from your solar panels to battery bank without any serious power loss. For example, we can say about flowing water through a pipe, if the size of the pipe is smaller, then very little amount of water can pass through it. Following steps are necessary for sizing wire for solar panels:

First Step:

First of all, you have to decide the required voltage for your system. Typically, it is 12, 24 and 48 volts. Solar panels need the required size of wire which can last long. The simple equation is that if the voltage is higher, then solar panels need the smaller size of wire in order to carry the current smoothly and safely. In the power equation (P=V x I) of a circuit, it has been observed that the Power (Wattage) "P" is equal to the Voltage "V" times the current I. Therefore, it is realized that if the voltage increases the current decreases, because V x I always equal to P. If, the total amount of current is very small. then you will need the wire which is small in size. Therefore, a higher system voltage is chosen as a thumb rule. You have to remember that all your equipment must run with your specified system. For example, if you choose the system voltage 24 volts, then your solar panels, battery bank, inverter and solar charge controller will require be 24 volts.

Second Step:

In this step, you have to determine the maximum currents (amps) are produced by your solar panels. This can be determine by multiplying the rating of one panel with the quantity of panels in your array. For example, if two 12 volt panels are connected in series to increase the voltage to 24 volts, then two panels should be count as one. This is done in this way, because in a series circuit, the current remains the same, but the voltage increases. For example, we can say about 12 solar panels rated at 12 volts and 6 amps. If you need the system voltage of 24 volt, then you should wire 2 panels in series to create required 24 volts for your system. You should do this 6 times. When 6 pairs are wired in parallel, 6 times 6 amps currents are added which provides you total current of 36 amps. This is the maximum amps your wires will carry.

Third Step:

Now, you need to determine the distance in feet from your solar panels to the solar charge controller and battery bank location. Never double the distance, even though indeed you will be running two wires, one negative and one positive.

Fourth Step:

Due to the resistance of the wire, there will be a transmission loss of the electrical power from your solar panels to your equipment location. You can't avoid this. Typically 3, 4 and 5 percent floss factors are considered for 12, 24, and 48 volt systems respectively. Based on copper wire using the standard AWG (American Wire Gauge) sizes, 00, 000, and 0000 gauges are generally referred as 2/0, 3/0 and 4/0. There are comparatively larger in size. A 4/0 size wire is fairly large. If you use this in a 48 volt system with a 5% loss factor, then you can expect the flow of current of 100 amps over 250 feet. This is considered as a very large system. Whatever gauge wire you use, you have to ensure that it is capable to carry the required amount of current produced by the system.

CHAPTER-9: What is Solar Charge Controller?

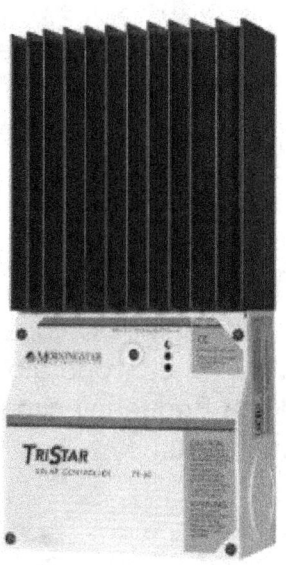

Solar charge controller is a device which regulates voltage and/or current to protect batteries from overcharging in the solar photovoltaic system. It has been observed that most solar panels of 12 volts have output voltage of 16 to 20 volts approximately. So, if there is no regulation the batteries will be damaged due to overcharging. In order to get fully charged, most batteries require around 14 to 14.5 volts. A solar charge controller regulates the rate of adding and subtracting of electric current from batteries. It helps to prevent batteries from overcharging and deep discharging. It is also protective against over voltage. This feature of solar charge controller contributes to enhance the lifespan of batteries significantly.

Types of Solar Charge Controller

Various shapes, sizes, features, and price ranges of solar charge controllers are available in the market. They are within the range of 4 amps to up to the 60 to 80 amp MPPT programmable controllers having computer interface. The full abbreviation of MPPT is Maximum Power Point Tracker. It's an electronic DC to DC converter which optimizes the match between the photovoltaic array (solar panels) and the battery bank or utility grid. They convert a higher voltage DC output from solar panels (and a few wind generators) down to the lower voltage required to charge batteries. If any system requires currents over 60 amps, then two or more 40 to 80 amp units are connected in parallel.

Basically, solar charge controllers are three types. They are as follows:

Relay based charge controllers: These types of charge controllers depend on relays or shunt transistors to control the voltage in one or two steps. When a specific voltage is reached, they necessarily disconnect the solar panel.

PWM charge controllers: PWM stands for Pulse Width Modulation. It is often used as float charging method. Using a very rapid "on-off" switch, it sends out a series of short charging pulses to the battery. The controller always checks the state of the battery to determine how quickly and how long (wide) send pulses should be sent. When the battery is fully charged with no load, it provides "tick" every few seconds and sends a short pulse to the battery. The controller may turn into "full on" mode or the pulse would be almost continuous when battery is discharged. The state of charge on the battery between pulses is checked by the controller and adjusts itself each time. These type of charge controllers now very much industry standard.

Maximum power point tracking (MPPT) charge controllers: MPPT charge controllers are high frequency DC to DC converters. They take DC input from the solar panels, turn it into high frequency AC and convert it back down to a different DC voltage and current for matching the panels to the batteries perfectly. MPPT's can operate at very high audio frequencies, usually in the range of 20-80 kHz. High frequency circuits are designed with very high efficiency transformers and small components.

Some linear (that is, non-digital) MPPT's charge controls are also available in the market. They are comparatively cheaper than the digital ones in terms of price and design. Somehow they can improve efficiency. But overall efficiency of these charge controllers can vary a lot and sometimes lose their "tracking point" when a cloud passed over the panel.

MPPT's are most effective under these conditions:

> ➢ Solar panels work better at cold temperatures, but without a MPPT you are losing most of that. Cold weather is most likely in winter - the time when sun hours are low and necessary power is required to recharge batteries most.

- MPPT puts more current into the battery when the state of charge is lower in the battery. This is another condition when the extra power is most required.
- When panels are 100 feet away and 12 volt battery is charging, the power and voltage drop are considerable unless very large wires are used. But, it's too much expensive. On the other hand, when four 12 volt panels are connected in series for getting 48 volts, there is very little power loss and the controller can convert that high voltage into 12 volts at the battery. That is, when a high voltage panel setup feeds the controller; comparatively smaller wires should be used.

Most controllers are designed by various type of indicator, a simple LED, a series of LED's or a digital meter. At present, built in computer interfaces are used in controllers for monitoring and control. The simplest one has a couple of small LED lamps which shows the available power and charge in the system. Both voltage and the current coming from the panels and the battery voltage are displayed in the meter of this charge controller. Amount of current is being pulled from the load terminals are also displayed in some charge controllers.

CHAPTER-10: What is inverter in solar PV system?

Inverter is a power converter which converts Direct Current (DC) into Alternating Current (AC). As most of modern appliances operate on 120 volts AC, an inverter plays a key role for your solar power system. It can convert the low voltage DC to the 120 volts AC and also can charge the batteries if connected to the utility grid.

Basically, an inverter can supply three types of power. They are as follows:

Usual or typical power: An inverter has to supply usual or typical power on a steady basis which is continuous rating. It is comparatively lower than the surge power. For example, after the first few seconds, a refrigerator pulls this power to start up the motor.

Surge or peak power: Supplying maximum power by an inverter for short time period is called surge or peak power. Typically, it ranges from few seconds up to 15 minutes or so. For example, we can say about electric motors like pumps which require higher startup surge while running.

Average power: This power is comparatively lower than typical or surge power. Typically, it is not any factor while choosing an inverter. While estimating required battery capacity, average power is only useful factor. Inverters need to be sized for the typical continuous and maximum peak load. For example, running a small television or a pump for 20 minutes during a one-hour period, the average should be approximately 300 watts only, even though the pump requires 2000.

Types of inverters:

There are various types of inverters are available in the market. Usually, the size of inverters are rated in the range from 50 watts up to 50,000 watts, although, sometimes units larger than 10,000 watts are used in solar photovoltaic systems or in household. Various types of inverters are as follows:

Square Wave Inverters: These types of inverters are not desirable at all. They produce inefficient square wave which is horrendous for running appliances. They are comparatively very cheap and the size of these inverters are typically 500 Watt or less. These types of inverters should not be considered for a home or solar photovoltaic system.

Modified Sine Wave Inverters: These are most economical and popular inverter. Usually, they produce an AC waveform in between a square wave and a pure sine wave. Modified Sine Wave Inverters are also called Quasi-Sine Wave inverters. They are not too expensive and perform well in almost all household appliances. Most computers perform well with a Modified Sine Wave inverter. But, it is not good for appliances which use timer or motor speed controls.

True Sine Wave Inverters: Among all inverters, a True Sine Wave inverter produces considerably a pure sine wave. These type of inverters are comparatively very expensive. Practically, it can run all kind of AC equipment perfectly. Most of True Sine Wave power inverters are controlled by computer. They can automatically turn on and off as per requirement of AC loads. If you need to supply automatic power to a normal home using a wide variety of electrical devices, it is recommended to use a True Sine Wave inverter. It has been observed that most appliances operate more efficiently and smoothly with a True Sine Wave inverter consuming less power.

Grid Tie Inverters: Grid Tie Inverters are suitable for the system which is connected to grid power supplied by utility company. Whatever electricity your solar panels produce, by using a grid power inverter, you can expect reduced electricity bill from your utility company. You can also sell back your excess power produced by solar panels to tour utility company. In this system, a much smaller battery bank should be installed in order to cover short term outages from a few minutes to an hour or so. If there is no frequent long term power outages and back-up power requirement, no batteries are required indeed.

Installing two inverters in a system is known as stacking which can provide more power or higher voltage. You can increase the output voltage by stacking two compatible inverters in series. This would be the technique to use to provide 120/240 volts AC. On the other hand, if you want to increase the power, you should configure them in parallel. For example, if two 2000 watt inverters are connected in parallel, then it will provide 4000 watts (4KW) of power.

CHAPTER-11: How to install solar panels?

In order to generate electricity for both commercial and home use, solar panels are used. For getting maximum possible sunlight and generate maximum electricity from the system, photovoltaic panels are installed on the roof top in both cases. Following steps are followed in the installation process:

First Step: Installing Mounting Structures of Solar Panels

In the first step, the mounts are required to be fixed for supporting the solar panels. It can be Roof-ground mounts or flush mounts which dependent on the requirement. This base structure provides support and strength. Precaution is taken on direction in which the PV panels (mono crystalline or poly crystalline) will be installed. The best direction to face solar panels is south for countries in the Northern Hemisphere because it gets maximum sunlight. East and West directions will also do. On the other hand, the best direction is north for countries in the Southern Hemisphere. The mounting structure needs to be slightly tilted. Angle of the tilt depends on the latitude of the location. Typically, it is in between 18 to 36 Degree. In order to increase the conversion efficiency, a solar tracker can be used.

Second Step: Solar Panels Installation

In the second step, solar panels are required to be fixed with mounting structures. By tightening nuts and bolts, the installation process can be done. Precaution is taken to secure the whole structure appropriately so that it is robust and persists extended.

Third Step: Electrical Wiring

In the third step, electrical wiring is done. Universal Connectors such as MC4 are used during wiring as these types of connectors can be connected with all type of solar panels. These panels must be electrically connected with each other in following series:

Series Connection: In this type of connection, the Positive (+) Wire is of one PV module is connected to the Negative (–) Wire of another module. This type of wiring increases the voltage match with the battery bank.

Parallel Connection: In this type of connection, Positive (+) to Positive (+) and Negative (–) to Negative (–) connection is done. In this type, wiring voltage of each panel remains same.

Fourth Step: Connecting the System to PV Inverter

In the fourth step, the system is required to be connected to a PV inverter. The Positive wire from the solar panel is connected to the Positive terminal of the inverter and the Negative wire is connected to the Negative terminal of the inverter. In order to generate electricity, the PV inverter is then connected to the Solar Battery and Grid input.

Fifth Step: Connecting PV Inverter and Solar Battery

In the fifth step, the PV inverter and the solar battery are required to be connected. The positive terminal of the battery is connected with the positive terminal of the inverter and negative to negative. In order to store electricity backup, battery is required in off grid solar system.

Sixth Step: Connect Solar Inverter to the Grid

In the sixth step, the inverter is required to be connected to the grid. For creating this connection, a normal plug is used to connect to the main power switch board. For supplying electricity to the home, an output wire is connected with electric board.

Seventh Step: Starting PV Inverter

After completing all the electrical wiring and connections, the PV inverter is required to be switched ON the Main Switch of the Home. Digital display of the PV inverter shows you stats regarding generation and usage of solar unit.

CHAPTER-12: What is grounding solar PV system?

In the context of safety, it is very important to ensure proper grounding in solar photovoltaic (PV) system. Basic PV module can produce potentially dangerous currents and voltages for the life of the system. Properly maintained grounding helps ensure the overall safety of the system. In electrical and PV systems, there are two types of grounding. They are: equipment grounding and system grounding.

Equipment Grounding:

Equipment grounding is also known as safety grounding or protective earthing. In this grounding system, all exposed non-current carrying metal parts of the electrical system are effectively bonds (electrically connects) to the earth (ground). By properly bonding exposed metal surfaces together and to the earth, the potential difference between earth and the conductive surface during a fault condition is reduced to near zero, reducing electric shock potential. The proper bonding to earth by the equipment grounding system is crucial, because most of the environment is at earth potential. The conductors used to bond the various exposed metal surfaces together are known as equipment grounding conductors (EGCs).

System Grounding:

When one of the circuits (current-carrying) conductors is bonded (connected) to the equipment grounding system and also to earth is called system grounding. The circuit conductor which is connected to the equipment grounding system and to earth is called the grounded conductor. The connection between the grounded conductor and the equipment grounding system is known as the system bonding jumper. Only one system bonding jumper is allowed in each separate electrical system in which the system grounded conductor is isolated from the grounded conductors of the source or other systems.

Earth Connection:

For making contact with the earth, a metallic device is used which is called the grounding electrode. The conductor which connects the central grounding point and a grounding electrode which is in contact with the earth is known as the grounding electrode conductor (GEC).

Solidly Grounded:

Solidly grounded is defined as connected to ground without inserting any resistor or impedance device in the circuit. In this case, fuses, circuit breakers, and mechanical relay contacts are used in certain grounding circuits.

Grounded Systems:

When one of the DC conductors (either positive or negative) is connected to the grounding system, which in turn is connected to the earth, then a PV system is defined as a grounded system. The conductor that is grounded basically depends on the PV module technology. Most modules can be used with a negative grounded conductor or even in an ungrounded system, but a few PV module technologies require the positive conductor to be connected to earth.

Color Codes:

The DC grounded conductor must have insulation colored white or gray or have three white stripes if it is 6 AWG (American wire gauge) or smaller. Larger size conductors must be marked with these colors at their termination points. Grounded PV source and PV output conductors 6 AWG and smaller are allowed to be marked in the same manner as larger conductors, in order to allow the use of the durable, black-insulated USE-2 and PV cable/PV wire in exposed locations within the PV array.

Grounding Electrode Conductor—Installation:

The connection between the DC grounded circuit conductor and the grounding system is usually made through the ground-fault protection device (GFPD) internal to most non-battery-based utility-interactive inverters in utility-interactive PV systems. It is important to make a provision for the internal connection of a DC GEC. Microinverters which ground one of the module circuit conductors must have an internal GFPD and will also require a GEC terminal. The connection at the inverter for the GEC is usually a marked GEC terminal. From that terminal on the listed inverter, the PV installer must make the connection to earth through a grounding electrode appropriately.

CHAPTER-13: What components are used in a solar mini grid project?

Basic technical components of a solar mini grid are grouped into three systems. They are: Production System, Distribution System and End User System.

Production System

A solar mini grid's overall capacity to provide electricity to end users is determined by the production system. Energy generation technologies, inverters, a management system and storage (batteries).

Energy Generation Technologies

Diesel generators, hydro power systems, solar photovoltaic (PV) modules, wind turbines, biomass-powered generators and geothermal-powered generators are included in a solar mini grid energy generation technology. It's a mix of sources (hybrid) of renewable or nonrenewable.

Inverters

An inverter can convert electrical current from DC to AC. Based on end user's requirement, power inverters are used in the system. Some energy generation technologies produce direct current (DC) while others produce alternating current (AC).

For example, a Solar PV module can generate only DC. Almost all mainstream appliances of households need AC. So, in order to provide required AC to those appliances of households, a solar-powered mini-grid serving needs inverters as part of its production system. On the other hand, battery charging needs DC power.

The energy generation and storage systems each have their own inverter in an AC-coupled configuration with storage (a battery). These separate inverters connect to one another on the AC side of the system. In order to control charging and discharging, battery inverter can be used. The energy generation and energy storage systems are shared in an inverter in a DC-coupled PV configuration. DC coupling can provide better performance; battery charging is more efficient when there are fewer power conversion steps.

Management System

Management systems are included in a solar mini-grid system. It can measure, monitor and control electrical loads. For example, we can say about a charge controller. To prevent the battery from charging and over charging, it is connected between the solar panel and the battery or inverter/charger. In the same way, metering and monitoring equipment allow mini-grid managers to gather data about energy use across end users, which informs operational decisions. In order to optimize performance, management systems often couple computerized energy management tools with smart metering. Some management systems allow operators to control the system remotely, including shedding loads as per requirement.

Storage

Energy storage (such as batteries) are required in some mini-grid production systems. For example, solar and wind resources are non-dispatchable. This means they only produce power when the renewable resource is available, not according to user demand. If end users require power on demand, the mini-grid must be able to store energy and supply it when resources are not available. Energy storage adds stability to the system by storing energy for peak consumption. Large mini-grid systems that run diesel generators continuously do not require batteries, but nearly all other mini-grid systems require some type of energy storage.

In order to optimize system performance, longevity and cost, project developers need to identify the most appropriate energy storage technology for their mini-grid. Lead-acid batteries are the most common, but fuel cells and advanced battery technologies like lithium-ion and sodium-ion batteries are generally more efficient and last longer. Costs for these new technologies continue to decrease. Large battery banks can cause safety hazards. High concentrations of hydrogen gas can cause explosions, and leaks can cause electrolyte spills. Batteries should locate in well-ventilated locations such as outbuildings or utility rooms.

Distribution System

Generated electric power is delivered from the energy production system to end users by the electricity distribution system. Distribution and/or transmission lines, transformers and the infrastructure to support the lines, such as poles are key components of a distribution system. Lines can be overhead or underground. Overhead transmission is most common as it is comparatively cheaper.

A variety of voltages are used in the distribution system. It can be either AC or DC and either single- or three-phase power. AC voltage levels in a mini-grid network covering a large area can be changed by transformers. To transmit electricity more efficiently over a distance, AC output voltage can be increased by step-up transformers. On the other hand, step-down transformers decrease the voltage from high- or medium-voltage transmission lines to 120 V or 220 V for residential use. Distribution network costs and system losses in AC mini-grids can be minimized by transformers. Distributing electricity at medium voltage allows systems to use smaller conductors, minimizing cable costs. Higher voltage causes greater safety risks for operators and users. Therefore, special trainings are required for operators.

There are various efficiencies in various components. Therefore, the determining of voltage, current and transformers impact energy losses. Cost usually dictates which option project developers choose. DC is generally less expensive than AC, because AC requires power conditioning equipment. The availability of appliances for different currents should also be considered by developers.

End user System

End-user systems provide an interface for end users to access, use and monitor electricity from the mini-grid is provided by the end user system. It takes into account consumers' needs and energy uses. For example, we can say about businesses require operating machinery for productive uses need different systems than households that use electricity for lighting and small appliances. The end-user system consists of connections to and from the mini-grid, systems to prevent electrical shocks and harm to both equipment and users and power consumption metering.

Energy consumption can be monitored in the end user system. It can estimate the cost of their consumption and understand the current status of the system. It can provide report on consumption rate and timing, like when and how much energy is used. Therefore, system operators can easily estimate and predict demand and consumption patterns. The data also allows regulators to establish tariffs that balance the needs of the operator and the consumer, while ensuring differing use cases are priced fairly and competitively. In order to ensure the safety of its users and protect valuable and expensive equipment, the system provides important electrical bond and grounding mechanisms.

Innovative metering and payment systems automate these otherwise complex tasks like making metering, billing and collection time consuming. The greatest degree of control over energy use can be provided by individual meters (one per end user). Meters can be pre- or post-paid; pre-paid meters typically are known as pay-as-you-go (PAYG) metering. Smart Meters are typically considered as newer generation meters. Although traditional meters or old generation meters are still in use. Smart meters do more than transmit payment and consumption data.

Utilizing mobile phone technology, smart meters gather data on energy consumption and facilitate two-way communication between the energy provider and end user. For example, a smart meter installed on the side of a residence communicates wirelessly with the utility systems. Some metering systems can log and report power quality and reliability data in real time, which helps to address the power quality issues before they become problems. Smart meters can even allow the grid operator to shut down troublesome loads before they jeopardize the entire system and cause problems for other consumers. Therefore, grid operators can ensure a higher quality of service to all customers by providing smart meters having these features.

CHAPTER-14: How to build a solar power plant?

As the global demand of electricity is increasing day by day, the necessity of establishing power plants is also increasing. Besides establishing conventional energy sources like oil, gas and coal based power plants, non-conventional energy sources like solar based power plants are also establishing gradually. Ground mounted solar power plants require feasibility, environmental assessment, grid connection, siting etc. which is typical for a large industrial project. Implementing a solar power plant project requires comparatively more time than a rooftop residential solar installation.

Countries having subsidy program like tax rebates, feed in tariffs etc. implement solar farms. Due to availability of solar panel, solar inverters and installation expertise, building a solar power plant is quite convenient at present. On the other hand, the permits and regulations for building a solar power plant or a solar farm vary from country to country and region to region based on federal and state laws. Before building a large scale solar power plant, lots of permitting steps need to be passed. These requirements are comparatively less for building a smaller solar power plants in the range of 5 MW. Basic steps required to build a Solar Power Plant are given below:

Selecting the site: Selecting a suitable site for building the solar power plant is crucial. The site should be free from shadow and trees. The Area must have easy access to the roads and the power grid.

Initial Financial Analysis: Initial financial feasibility with information like the solar insolation, land costs and interconnection possibility with the power grid operator.

Land Acquisition: In this step, land acquisition starts through the transfer of ownership and leasing from government.

Engineering Design/Selection of the Technology: An Engineering Layout is prepared along with the Selection of the Technology and Vendors of Solar Equipment.

Permitting: In this step, various permitting procedures should be followed. This is specific to an area and can be quite cumbersome. Permitting involves very heavy costs forming almost 15-20% of the cost of a Solar Power Plant Project.

Power Purchase Agreement: A Power Purchase Agreement (PPA) required to be signed with the Power Utility who will buy the Electricity.

EPC Selection: A Solar EPC Contractor or a System Integrator is selected. In case an EPC Contractor is selected, then Solar Panels, Mounting and Inverters needs to be purchased if the Contract is not a Turnkey One.

Financing: Financing of the Solar Power Plant Project should be done. Solar Power Plants require a high initial investment with very low operation and maintenance costs. Typically, 60-80% of the Project is Debt Financed.

Testing and Connection to Grid: After building the Solar Power Plants, testing of the plant must be done before connecting to the Power Grid.

Ongoing Operation & Maintenance: A Solar Power Plant has a life span of 25-30 years and requires less maintenance and monitoring. Solar Inverters should be replaced after 10-15 years. In case a Solar Module fails, it needs to be replaced as it degrades the performance of other Solar Panels.

Building a solar power plant requires project management and technical skills distinctive of an industrial project with some unique features. The above steps are a simplistic procedure for building a solar power plant. The process in fact requires in depth and concrete execution skills. A solar power plant can be built in 3 months to 2 years depending on the permitting and the expertise required.

CHAPTER-15: How to maintain solar panels?

 Solar panels play a significant role for generating electricity. Solar panels generate electricity when exposed to the sunlight. The efficiency of solar panels and generating electricity is strongly influenced when sun exposure to the solar cell is reduced. Dust, dry areas and building up other particles like droppings of birds have significant impact on the amount of generating electricity by solar panels. Cleaning solar panels regularly can significantly increase the overall efficiency of solar panels. Typically, solar panels are self-cleaning. Insufficient care for solar panels can decrease the efficiency of solar panels. Therefore, it is very important to maintain solar panels very carefully.

How to clean solar panel glass

➢ Before starting cleaning, you have to shut down your entire system as per instructions or guidelines of your manual provided by your solar panels manufacturer.
➢ Cleaning solar panels from the ground is safe. In order to stay on the ground safely, you need a good quality squeegee with a plastic blade and soft brush on one end and a cloth covered sponge on the other coupled having a long extension. For reaching water stream to panels, you should use a hose with a suitable nozzle. If it is not possible to clean panels from the ground, never attempt to access your rooftop unless you have necessary training and relevant safety equipment. If you don't have any training, you should complete the cleaning process by a suitably qualified professional.

➤ You should clean your solar panels early morning or in the evening and on an overcast day. Using water while beating down the sun on solar panels can quickly evaporate which causes smearing dirt. Falling dew on solar panels overnight makes dirt soften which helps you to clean panels using less water and energy. So, typically it is ideal time to clean panels early morning.

➤ For removing slab on materials, never use harsh abrasive products or metal objects. Scratching the glass on a solar panel spreads shadow which decreases the performance of a solar panel. Don't use corrosive powder as it risks scratching the panels. Using detergents may smear the glass of a solar panel. So, it should be avoided.

➤ If your solar panels are dry, you should dismantle all loose materials before approaching panels with water. It will help you to clean quickly and easily.

➤ In some installation scenarios, oily splash may occur. For example, if you live adjacent to and downwind of a major roadway wherein vehicles like trucks move most frequently or you live near an airport. If you notice appearing oily splash on your panels, then you should use isopropanol as a spot-cleaning element.

➤ You can easily remove the most persistent dirt exists on the glass of a good quality solar panel by washing gently with a soft brush or plastic cleanser or a prickly cloth covered sponge using clean water.

➤ Use rainwater and mineral-rich water as a final rinse, after then you should compress dry. If your available water is hard or mineral-rich, you must compress well, because mineral-rich water can form deposits on glass as it dries.

If you want to obtain the highest efficiency possible, then you need to install a special 'PID (Potential Induced Degradation) coating' on the glass of your solar panels. This may be in the form of a spray or by mixing it with the water for the maintenance of your solar panels. In this way a filter exists on the panels, so that the electric charge quickly ends up in your installation. It should be noted here that the PID coating works properly only when a PID box is installed appropriately.

CHAPTER-16: How to troubleshoot a solar PV system?

We expect smooth power generation from a solar photovoltaic (PV) system while sun shining. When the sun is out but the system doesn't generate electricity as per required capacity, then we consider the system as problematic. Solar photovoltaic (PV) system occasionally requires troubleshooting as like as other energy systems. The key to your success depends on the strategy you choose for identifying the trouble source. Typically, three problems occur in solar photovoltaic (PV) system. They are-an array problem, inverter problem, or load problem. Concentrating on common troubleshooting problems and solutions can ensure that your system is taking the advantage of summer's sunny days.

Components of a solar photovoltaic (PV) system

Cell, module and array: A typical photovoltaic (PV) cell produces a small electrical output ranges from 0.5W to 2W approximately. As these devices are electrical, we can boost overall output level by wiring them in series and parallel strings. Wiring PV cells in this method is called a module. Some manufacturers now manufacture "power modules," which can generate 190W or more power. Under full sun shine conditions, a typical 190W module which is connected to a load might generate the voltage and current of approximately 27V and 7A respectively. When modules are wired in series and parallel strings, then it is called an "array." The output of an array is designed in such a way that the output of an array can meet almost any electrical requirement of small or large scale system.

Combiner box: Desired voltage and current can be gained by wiring modules into an electrical string. All strings are combined into one electrical output in the combiner box which is then fed to the inverter.

Inverter: Inverter converts the DC output into AC as per requirement of any photovoltaic (PV) system. Electric utility grid connected inverters generate AC which is identical to the power generated by the electric utility. These inverters sense the waveform characteristics and generated voltage of the electric utility and generate the same type of AC.

How to troubleshoot the problem of an array?

The input voltage and current level of the inverter's from the array need to be checked and recorded. If the array is not generating required DC electricity, check all switches, fuses, and circuit breakers. Blown fuses should be replaced and the breakers and switches need to be reset. A specious surge might have passed through, blowing or tripping the protective devices. Loose or dirty connections and broken wires in the inverter should be checked. All connections should be clean and make them tighten. All damaged wires should be replaced. Check the array visually for obvious damage to the panels and wiring.

There are fuses for each module or sub-array string in many combiner boxes. While troubleshooting, all these fuses should be removed and the current reading and open-circuit voltage should be recorded for each circuit string. Low output voltage indicates that some panels in the series string are disconnected or defected which requires replacement. Defective bypass or blocking diodes in the modules might need to be replaced. Wrong wiring connecting the modules in the string to the combiner box, junction box or the inverter may cause low voltage. Undersized wire may cause this problem. This problem should be rectified by upgrading the wire size for the current level.

During overcast or cloudy conditions, a damaged panel or defective bypass diode can produce low output current. One or more parallel connections between modules in the string might be loose, broken, or dirty — or some parallel connections in the module might be loose, broken, or dirty. Replace a damaged module or one with internal parallel connection problems. Defective diodes should be replaced. All connections need to be tightened and clean. Shades on the array decreases output current significantly. So, in order to obtain full current output from the string, it is required to remove the source of shades from array.

Dirty modules decrease the output current. In order to restore the output current of array, these modules need to be washed. After washing, the output current should be checked again.

How to troubleshoot the problem of an inverter?

First of all, operating DC input voltage and current level of the inverter should be checked by a volt meter and DC ammeter respectively. After then, these data should be recorded. On the AC side, check the inverter's output voltage and current level. A blown fuse, broken wires or a tripped breaker can cause the insufficient output power from the inverter.

Some inverters have LED displays as indicators. It is necessary to check whether these LEDs are blinking properly or not. Properly blinking LEDs should indicate the actual condition of the inverter.

True-rms reading type volt meter can be used to measure the voltage and current to measure. After measuring, the kilowatt output should be recorded. You should record the total kilowatt hours generated since it first started up which is displayed by the inverter. You can compare the PV system's production since the last inspection by using the recorded data.

AC load side of the inverter should be measured, because load on the inverter might have too high demand of a current. In this scenario, inverter should be upgraded or loads should be reduced.

Before starting the inverter again, any ground faults should be checked and repaired after shutting down the power. Inverter can sense the voltage and frequency of the electric utility. Typically, it generates AC electricity at the same voltage and frequency. The AC output current output of the inverter fluctuates with the level of solar input on the array. If internal disconnects sense that the electric utility voltage is high or low, then it will shut down the inverter. If this problem remains, then you should contact the concerned authority of electric utility to rectify the problem. Inverter problems could also be caused by a problem on the array side of the inverter, which trips one of the internal disconnects.

How to troubleshoot a load problem?

All load switches should be checked first. It should be checked whether they are turned off or placed in the wrong position. You should check and ensure that the load is plugged in. Next, the fuses and circuit breakers should be checked. If there are tripped breakers or blown fuses, the cause should be located and the faulty component should be replaced or fixed. If there are no tripped breakers or blown fuses and the load is a motor, then there might be an open circuit in the motor or an internal thermal breaker might be tripped. In this scenario, you need to plug in another load and observe its operation.

Any loose connections and broken wires should be checked. All bad wiring should be replaced. After shutting down the power, any ground faults should be checked and repaired. Fuses need to be replaced and the switches should be reset. If they blow or trip again, there might have short circuit problem, which must be located and repaired.

If the load does not operate properly, the system voltage should be checked at the load's connection point. Too small or too long wire feeding the circuit may cause the low voltage which needs to be upgraded to reduce the voltage drop. The load might also be large enough for the wire size in the circuit. In this scenario, the size of wire need to be upgraded or the load on the circuit should be reduced

CHAPTER-17: What are the challenges in solar power generation?

One of the most promising renewable energy technologies is solar power which allows generating electricity from free, unlimited sunlight. It's one of the rapid growth industry and technology which contributes significantly to homeowners and commercial industry. It would be a reasonable and useful solution for new power generation installations in developing countries to be powered by carbon emission free sources like solar. Solar seems perfect for countries with lots of sun exposure and no efficient way of bringing the conventional power grid to remote locations. But solar still faces a number of obstacles before it can really replace fossil fuels for power generation. However, there are many unexpected challenges with solar electrification that entrepreneurs are learning about while doing business in these developing countries.

The Levelized Cost of Solar Power

Levelized cost of Solar Power is a term which describes the cost of the power produced by solar over a period of time, typically the warranted life of the system. It is the capital cost for solar power plants which is high initially. Moreover, there are on-going maintenance costs for both types and the cost of financing any loans.

Intensity of Solar Radiation

Intensity of solar radiation is one of the main obstacles to the extensive application of solar power. It varies with different locations of the world. The amount of useful solar energy incident in any particular location is highly dependent on latitude and climate. The equator receives the most annual solar energy and the poles receive the least. Dry climates receive more solar energy than those with cloud cover.

Required Land Space

A good amount of land space is required for solar farms, as power generation is directly proportional to the surface area covered. Therefore, the largest solar farms in the world are built in deserts and huge open spaces. However, this is not feasible in smaller countries with limited landmass, or even for larger countries where a land compromise develops restricting the use of agricultural land for developing solar farm.

Transmission

Sufficient transmission is required to transport the power to urban load centres. Intermittent resources like solar can pose distinctive problems in transmission planning and in efficient operation of transmission infrastructure, causing in higher transmission costs, increased congestion, and even generation limitations when sufficient transmission capacity is not available. Due to potential transmission barriers, solar project developers will need to evaluate the economic trade-off of locating where the resource is best versus locating nearer to loads where transmission barriers are less possible.

Reliability

Reliability is one major problem with solar power. A solar panel can produce electricity for maximum 12 hours a day and a panel can only reach peak output for a short period around noon. Solar panels with tracker can track the sun spreading the major generation period fairly, but it still means that panels employ very little of the day producing at maximum capacity. During peak generation, storage batteries can be charged by solar panels which help to supply a dribble of power at night. But they can be costly, contain toxic materials and wear out rapidly due to frequent charging and discharging cycles.

Efficiency of Solar Panel

Photovoltaic efficiency is another obstacle. In the desert area, a single square meter of solar panel could receive the equivalent of more than 6 kilowatt-hours of energy in the course of a single day. But a solar panel cannot convert that much of energy to electricity. The efficiency of a solar panel regulates usable power. Most commercial solar panels have efficiency less than 25%. The more efficient a panel is, the more expensive it is to produce.

Environmental Issues

Though generating power from solar is free from carbon emission, manufacturing of solar panels and associated technologies can comprise some environmentally unfriendly elements. Nitrogen trifluoride is a common by product of electronics manufacture; including those used in solar cells, and it is a greenhouse gas 17,000 times more potent than carbon dioxide. Moreover, many solar cells contain small amounts of the toxic metal cadmium, and the batteries required to store generated electricity can contain a host of other heavy metals and dangerous substances. As solar technology improves, manufacturers may be able to move away from these potentially dangerous substances, but for now, they ruin the otherwise notable environmental benefits solar power offers.